Snowy Owl

by Grace Hansen

Abdo Kids Jumbo is an Imprint of Abdo Kids
abdobooks.com

abdobooks.com

Published by Abdo Kids, a division of ABDO, P.O. Box 398166, Minneapolis, Minnesota 55439.
Copyright © 2020 by Abdo Consulting Group, Inc. International copyrights reserved in all countries.
No part of this book may be reproduced in any form without written permission from the publisher.
Abdo Kids Jumbo™ is a trademark and logo of Abdo Kids.

Printed in the United States of America, North Mankato, Minnesota.

102019

012020

Photo Credits: Alamy, iStock, Minden Pictures, Shutterstock, ©Peter Trimming p9/CC BY 2.0

Production Contributors: Teddy Borth, Jennie Forsberg, Grace Hansen
Design Contributors: Dorothy Toth, Pakou Moua

Library of Congress Control Number: 2019941220
Publisher's Cataloging-in-Publication Data

Names: Hansen, Grace, author.

Title: Snowy owl / by Grace Hansen

Description: Minneapolis, Minnesota : Abdo Kids, 2020 | Series: Arctic animals | Includes online
    resources and index.

Identifiers: ISBN 9781532188909 (lib. bdg.) | ISBN 9781532189395 (ebook) | ISBN 9781098200374
    (Read-to-Me ebook)

Subjects: LCSH: Snowy owl--Juvenile literature. | Owls--Juvenile literature. | Nocturnal birds--Juvenile
    literature. | Zoology--Arctic regions--Juvenile literature. | Birds--Behavior--Arctic regions--Juvenile
    literature. | Arctic--Juvenile literature.

Classification: DDC 598.97--dc23

Table of Contents

The Arctic

The Arctic is the northernmost part of Earth. It is made up of land, the Arctic Ocean, and the sea ice that floats on it. The weather there is freezing cold. Any animal that lives in the Arctic is tough!

4

Snowy Owls

Snowy owls like wide open spaces and few trees. This makes the Arctic their perfect home.

Snowy owls are some of the heaviest owls. They can weigh up to 6.5 pounds (3 kg).

Much of this weight comes from their thick, white feathers. These keep them warm in the freezing cold temperatures.

Young male snowy owls have markings on their feathers. Older males may be completely white. Females are covered in dark markings for their entire lives.

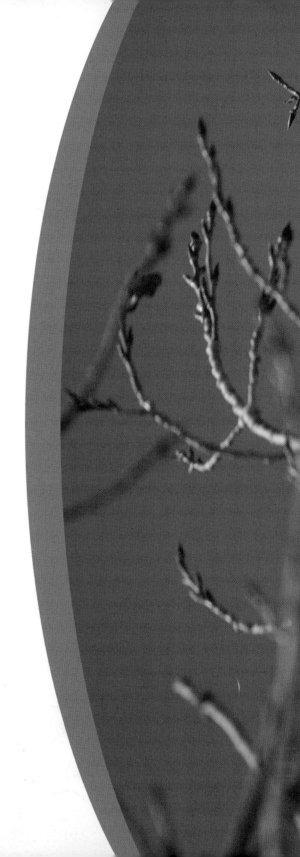

Sharp eyesight and hearing make them great hunters. They sit, watch, and listen for their **prey**. They will sit in the same spot for hours if they must.

Snowy owls capture their prey from above. They catch prey with their sharp talons and eat it in one gulp. The owls' favorite food is lemmings.

Baby Snowy Owls

Females build nests on the ground. They lay 3 to 11 eggs at a time. Mother owls sit on their eggs until they hatch. Males bring females food during this time.

Eggs hatch after about one month. Baby snowy owls are covered in very soft, white **down**. Their first feathers are gray. By 2 months old, they can already fly!

More Facts

- Snowy owls have round, bright yellow eyes. However, their eyes are smaller than other kinds of owls.

- Their eyes are likely smaller because they do their hunting during the day. Other owls often hunt at night.

- In the Harry Potter series, Harry has a snowy owl named Hedwig. In the movies, seven different male owls played the role. However, in the books Hedwig is a female owl.

Glossary

down – fine, soft, fuzzy feathers that cover young birds and are found underneath the outside feathers of some adult birds.

lemming – a very small mammal that looks like a mouse with a short tail.

prey – an animal that is hunted and eaten by another animal.

sea ice – frozen ocean water that is typically covered with snow.

talon – the claw of a bird.

Index

Abdo Kids
ONLINE
FREE! ONLINE MULTIMEDIA RESOURCES

Visit **abdokids.com** to access crafts, games, videos, and more!

Use Abdo Kids code

ASK8909

or scan this QR code!